A Fistful of Stars
Communing with the Cosmos

———•———

A Fistful of Stars

Communuing with the Cosmos

---•---

Gail Collins-Ranadive

with Milt Hetrick

THE LITTLE BOUND BOOKS ESSAY SERIES
Personal. Poignant. Powerful.

LITTLE BOUND BOOKS
Small Books. Big Impact | www.littleboundbooks.com

© 2018 • Text by Gail Collins-Ranadive
All rights reserved. Except for brief quotations in critical articles or reviews, no part of this book may be reproduced without prior written permission from the publisher: Homebound Publications, Postal Box 1442, Pawcatuck, CT 06379. WWW.HOMEBOUNDPUBLICATIONS.COM

The author has tried to recreate events, locales and conversations from her memories of them. In order to maintain their anonymity in some instances she has changed the names of individuals and places, she may have changed some identifying characteristics and details such as physical properties, occupations and places of residence.

Published in 2018 • Little Bound Books
Imprint of Homebound Publications
Front Cover Image © by Joseph Castells
Cover and Interior Designed • Leslie M. Browning
ISBN • 978-1-947003-97-2
First Edition Trade Paperback

10 9 8 7 6 5 4 3 2 1

Homebound Publications is committed to ecological stewardship. We greatly value the natural environment and invest in environmental conservation. Our books are printed on paper with chain of custody certification from the Forest Stewardship Council, Sustainable Forestry Initiative, and the Program for the Endorsement of Forest Certification.

This little book is dedicated to the stars, those we can't touch and those we can, including the human 'stars' whose insights have touched us: may they continue to illumine our minds, our hearts, our souls, and our lives.

ALSO BY THE AUTHOR

*Finding the Voice Inside:
Writing As a Spiritual Quest for Women*

Light Year: A Seasonal Primer for Spiritual Focus

Inner Canyon: Where Deep Time Meets Sacred Space

Chewing Sand: An Eco-Spiritual Taste of the Mojave Desert

Nature's Calling: The Grace of Place

Star Point One

"May I show you something?"

It was snowing cotton on the summer solstice when jogging Jim stopped me, Ancient Mariner-like, with his question.

We had been passing one another for several summers in a row, nodding our good mornings, but these were the first words that had ever passed between us. I followed him to the edge of the Highline Canal Trail where, under a canopy of cottonwood trees, he picked up a wind-blown twig and snapped it in two.

There, at the center of each end, lay a perfect five-pointed star.

Answering the question in my eyes, Jim shared that his children had come across this phenomenon one winter solstice while playing in the snow. They'd decided that the indigenous peoples who once lived here also knew that the cottonwood trees are full of stars!

I mumbled that modern science has proven what ancient people knew intuitively: everything on earth is made up of star stuff, us included!

I took the twigs with their tiny stars home, where an aging cottonwood tree presides over the backyard. A welcome sight in the treeless west, cottonwoods thrive near water sources, and pointed the pioneers towards rare streams hiding beneath their canopy of green. The broad leaves hold onto the fading fall sunlight and give it back as a yellow crown of gold. But it is the late spring, early summer reproductive cycle that defines this tree: wisps of cotton drift through the air and settle over everything, startling an observer with the delight of out of season 'snow.' And now here was yet another thing to appreciate about this amazing tree…its mysterious stars!

Coincidentally, the cottonwood in our backyard shades the very room in which my partner and I had created a mural of the newly unfolding story of the universe, whose fourteen billion-plus year origin has 'come to light' within our lifetime.

My partner Milt, a retired aerospace engineer who was part of the design team whose rockets sent humans to the moon and Cassini to Saturn, has long been immersed in the 'what' of this new narrative.

A retired liberal religious minister, I have always concluded funerals with "from ashes to ashes, dust to dust, star stuff to star stuff," thus hinting at the 'so what' of this new cosmology, a secular story with sacred dimensions. For physics teaches that every wave of every particle that was the deceased person remains with us in this world.

All the photons that ever bounced off her face, all the particles whose paths were interrupted by his smile, have raced off like children at recess, forever changed, to go on forever.

For in death, the atoms that composed that person have merely been repurposed. All energy, which originated during the Big Bang, will always be around. Thus a person's 'light,' the essence of her or his energy, will continue to echo throughout space until the end of time.

Yet death begs the question: what is the purpose of life? What do we make of our time between star stuff to star stuff?

Humans need to make meaning, discern the 'so what.' This knowing that we were fashioned in the stars aches to inspire and inform how we live here on earth. It longs to be translated into how we live together in community, how we understand ourselves, and how we connect with the ground of Being, the source of All Becoming…..

I stare down at the twigs in my lap and suddenly wonder if their stars can become as a microcosm of the macrocosm of which we're but a part.

For if we can poetically see the world in a grain of sand and heaven in a wildflower, why not the cosmos in a cottonwood twig?

The poet Robert Frost claimed that the person without the organizing metaphor for life was the truly lost soul. Perhaps this tiny star at the center of a tree branch can become such a soul-organizing metaphor.

For as Thomas Berry put it in *The Great Work*, "Our sense of who we are and what our role is must begin where the universe begins."

Our mural begins at that beginning.

The first wall includes a panel of images from the Hubble Telescope, an amazing piece of technology that Milt's colleagues helped develop so we can see back in time and into the outer reaches of the Universe.

Hubble himself could barely believe what his own research showed and what Einstein had doctored to deny in his own equations: we live in an expanding universe! And by starting with what is observable now, science could rewind Time and travel back to its very beginning.

Thus the images on our wall start with the explosion of light at the beginning of Time and show the first seven billion years of the fourteen billion years of a narrative we now call the Universe Story.

The second wall depicts the four billion years of Earth's story.

The third wall is for the ongoing human story.

From the vantage point of my favorite chair in the living room, I can focus on all that went on before our Mother Star gave birth to our sun.

Today's Universe Story tells us that, in the beginning, an unimaginable source of light energy burst forth. As it expanded and cooled, energy was transformed into matter, fundamental particles began to appear and then protons and neutrons and electrons emerged—the building blocks of atoms. Within a cosmological blink (300,000 years), all the hydrogen (one proton, one neutron, one electron) in the Universe emerged.

Out of clouds of hydrogen came galaxies filled with first and second-generation stars, including our own 'Mother Star.'

After fusing hydrogen into helium that fused into carbon that fused into nitrogen that fused into oxygen on through to iron, this supernova exploded into a cosmic mess out of which our sun and its planets formed.

By then this star stuff included all the other elements in the Periodic Table. The gravity of the iron-cored Earth captured the elements needed for our just-right planet to use as building blocks for the creation of life.

Today, our Mother Star shows her face in starfish and on sand dollars, in the star shape of flower petals and on the inside of fruit, as well as in human language: when we speak of disaster we mean we are 'separated from the stars.' Somehow we know in our very bones that we are part of the whole Universe! But more, as Thomas Berry has put it: "the galactic evolutionary process of the universe, the geobiological evolutionary processes of the Earth, and the cultural evolutionary processes of the human need to be understood, and celebrated as three components of the single evolutionary narrative."

What might it 'look like' if we were to embed our personal narratives within the context of the emergent story of the cosmos, a story in which mystic wisdom and quantum physics converge?

And what might be expected of us in this new era of consciousness?

Star Point Two

On the second wall of our dining room, the four billion year story of Earth unfolds. Formed from the debris of the explosion triggered when our Mother Star fused iron, our planet embraces it at its molten core. But our rock red planet had to turn watery blue before life could wiggle forth.

Lightning-zapped remnants of carbon from our Mother Star transformed into pre-DNA and began trying on a diversity of life forms. Cell biologist Ursula Goodenough describes this process as the emergence of 'something more from nothing but.' For instance, water is nothing but the combination of hydrogen and oxygen, yet becomes more than either.

And while humans may be but a clever way that water found for moving beyond the reach of ponds as Loren Eiseley claims, it is only 'we the people' who have been able to consciously reflect upon the whole of the evolutionary journey.

Thus Milt and I personified Earth's timeline by using the human body:

Your feet planted on the ground mark the beginnings of earth at the birth of our solar system 4.5 billion years ago, the first living cell appears at your ankle, life's common ancestor is at calf level, multiple cellular life emerges at your knee, DNA exchange comes into being at hip level, plants and oxygen happen at your shoulder.

Then, when you raise up your arms the largest explosion of life is at your elbow, dinosaurs come and go along your forearm, thus opening a bigger niche for those pesky little furry mammals darting between their feet. At your wrist is when hominids appeared. Our Earth wall turns the corner just as hominids became human.

The third wall depicts human development, from the taming of fire to the migrations across the planet, from drawing on cave walls to the invention of agriculture, and then on to the settling down into cities, states, and nations. Human culture is what made that journey possible.

Each of us repeats this evolutionary life-process, beginning when we experience our very own personal big bang event: that moment when an egg and a sperm explode into new possibility. We suddenly burst into being as something that has never existed before. But it will take nine months to become even a half-baked human.

Born before its brain grows too big for its head to pass through the birth canal, the human infant lacks the instinct and imprinting needed for survival. Thus we are born into a second womb: the human family.

By the time my second child was born, I was a stay-at-home mom and got to watch her 'climb the human family tree'…from being an out of water fish to a crawling amphibian to a clinging primate to a human choosing to stand up and walk, thus freeing her hands for new uses.

She did this all within the context of our little 'nuclear' family. Yet because the family unit is embedded in a particular culture, the society into which she was born became the primary container for shaping her beliefs and behaviors. But she also had a second family, an extended one living on the other side of the globe, in India.

Lewis Mumford claims that, because our human brains are initially so unformed and unfinished, outward societal rites and rituals must create a container to hold everything and everyone together for the greater good.

Living in two cultures teaches a child that there are different ways of doing the same human things. And the stories that reinforce these life-ways form the often-unconscious foundational narratives of each culture.

In my first published work, a picture book that featured this daughter on the cover, I invited children in my older daughter's first-grade class to imagine that, instead of being born in America, they had been born in India. They would grow up sleeping on the floor in a common room, eating with their fingers in the polite and proper way, playing on the flat roof of their apartment building in an overcrowded city, and never, ever wearing their outside shoes inside the house....things I wish I'd known about and prepared for ahead of time. These outward customs rested on religious convictions, such as respect for the Hindu god image kept in the kitchen, where only certain foods would be prepared on my mother-in-law's fast days, reminding me of eating fish on Fridays in the home of my Catholic aunt. My exposure to Hinduism's inclusively cosmological world-view expanded my fundamentalist Christian one, and would eventually lead me into the Unitarian Universalist ministry some twenty years into the future.

Meanwhile, during my first visit with my in-laws in Bombay (now Mumbai), my father-in-law wanted me to meet with a family friend who, as an amateur in astrology, had read his son's horoscope years earlier.

This astrologer had predicted things that were clearly coming to pass now: his becoming a doctor who would never practice general medicine, leaving India and never coming back there to live, marrying outside of his nationality, religion, and caste. (In an attempt to thwart this troublesome trajectory, my in-laws tried to get their son married off before he left India to continue his medical education in the United States.)

Although I was clearly part of the fulfillment of this prophecy, I refused to meet with the astrologer. Perhaps because I'd been born in the United States, I preferred to believe that my destiny was in my own hands, not somehow predetermined in the stars.

But was it really? After all, my 'second womb' was a family living in a Veteran's housing project and presided over by under-educated parents in an era when women stayed home having babies. Because there was no legal access to family planning, I was 'destined' to become the oldest of eight children at a time when the single wage earner's hard-earned savings weren't to be wasted on college for daughters, for starters. I managed to go to nursing school on an academic scholarship, and stay there when I received more funding through an award for the student showing the most promise after the first six months, presented at our 'capping' ceremony. It was while I was a student nurse that I met and then married the intern from India…..all on my own…or so I'd thought.

Although our shared name translated as 'glory of the battlefield,' it still came as a shock when my mild-mannered husband suddenly got drafted into the United States Army. He was not a U.S. citizen, but the Vietnam War had created a shortage of doctors and there was a pool of foreign medical graduates, none eligible to vote, just waiting to be tapped into by opportunistic politicians.

He was given three options: accept the invitation to serve in the army, leave the country and never come back, or go to jail. He said yes to military service, which, it turns out, was 'preordained' not only in his name but by his family's warrior caste: serving in the army for twenty-three years came 'naturally.' Natural for him perhaps, but not for me.

I'd descended from a pacifist couple who, having fled Germany during the Thirty Years War, then left their Pennsylvania farm during the American Revolution so that their seven sons couldn't be conscripted to fight. Eerily, I did not know that story while I was struggling to be a 'good' military spouse. But somehow I knew that war was never a long-term solution to conflict, especially with the development of nuclear weapons during my lifetime. And concern over putting my neighbors in harm's way propelled me into peace studies. I would eventually become part of a grassroots lobbying campaign to establish a federally funded institution that would explore more peaceful options for international conflicts.

It was after The United States Institute of Peace was funded by an amendment to the Reagan defense budget that I learned of my family's history. Was I just following a path that was 'written in the stars' for me?

I don't know. I do know that when a commanding officer 'counseled' me that war is just part of human nature, everything inside me screamed: speak for yourself! I belonged to a historical movement oriented towards peace, and in fact first met Milt during that national peace academy effort.

One of my fellow board members on that citizen's campaign was a Quaker sociologist whose research revealed that in prehistoric cultures cooperation was normative for a longer human time than the ruthless competition our current culture has been following for a mere few thousand years. Nowadays, that competitive narrative has ceased to be a nurturing womb, and become a toxic tomb that begs to be transcended.

Yet resistance to any such change seems relentless and absolute. Those with vested personal, professional, political, and/or financial interest in maintaining the status quo never have and never will give up their power without a struggle. Remember Galileo, terrorized with being burnt at the stake for his truth-claim that the earth moved around the sun, rather than the sun around the earth. To make that cosmological shift would have meant a reordering of societal norms that threatened the position of those in power. Thus they were determined to repress any and all revolutionary knowing that was 'dawning' through the human mind.

But we people seem 'hard-wired' to opt for growth and change! As both toddlers and teens we practice saying NO to the norms that seek to define and confine us. When cultural patterning becomes too rigid and/or when a particular belief system decides it contains the only truth for all time, and the culture begins to collapse in upon itself, some people will attempt the move into something new. They somehow sense that if things are not expanding and progressing, they're imploding and regressing, a message that seems to come from our own Mother Star!

Nowadays we're in the liminal space between stories. While this can be disorientating, and distressful, it can also be a deeply creative place.

But as humans we have to choose to engage in this process consciously. Yet it is this very consciousness that is growing in complexity and compelling us to reboot our beliefs and behaviors!

For as Lewis Mumford assures us in his classic *The Myth of the Machine,* "The establishment of human identity is no modern problem. Man had to learn to be human, just as he had to learn to talk, and the jump from animalhood to humanhood, definite but gradual and undated, indeed still unfinished, came through endless efforts to shape and reshape himself."

Star Point Three

At my twentieth high school reunion, a classmate backed me into a corner and began yelling at me. Apparently I had unwittingly snubbed him while I was head majorette and he was a band member, and he had been carrying a grudge against me ever since. Taken aback. I stood there stunned while he went on and on about what a terrible snob I was.

Yet I had been a painfully shy teen just trying to stay off everyone's radar screen. Had my awkward silence provoked some unconscious projection and provided a screen for this person to see in me what he couldn't accept in himself? For he too was reserved, and had rarely, if ever, spoken TO ME. What strange dance had we unwittingly engaged in?

The reason I was even a majorette at all was because I knew I wasn't pretty and perky enough to make the cheerleading squad. I didn't play an instrument, so the only other option was twirling a baton. I could learn how to do that, and then practice to perfection.

Through this strategy I was playing out the conversation I'd overheard my parents having late one night: they had observed that my younger sister fit naturally into activities, while I had to push myself into them.

While they were making an observation rather than passing judgment, I spent half a lifetime agonizing and self-judging and feeling guilty because I was not outgoing, friendly, and spontaneous. What was wrong with me?!

From Socrates to Emerson, we've been told that the unexamined life is not worth living. But Robert Fulghum (*All I Really Need to Know I Learned in Kindergarten*) goes on to warn the examined life is no picnic.

But alas, self-knowledge IS the ultimate hallmark of humanity, a phenomena that began back in the Pleistocene Era when, as paleontologist-priest Teilhard de Chardin observed, something within humans "turned back on itself and, so to speak, took an infinite leap forward." The Threshold of Reflection had been breached: "consciousness was now leaping and boiling in a space of super-sensory relationships and representations" and was simultaneously capable of perceiving itself.

Still, I couldn't 'go there' until I came across the insights and ideas of the Swiss psychiatrist Carl Jung. Through studying with one of his original pupils, I could understand, if not actually accept, my inherent personality.

I've come to accept that it's all about energy flow.

Our Life requires Light to energize it. Green plants photosynthesize the sun's light and turn it into the carbohydrates that are consumed by animals and humans. These carbs then power our bodies and big brains. As Loren Eiseley noted, we humans couldn't come into being until flowering plants had concentrated the sun's energy into protein-rich nuts and seeds and fruits: we simply couldn't have eaten enough grass to fuel the high metabolism needed to develop into what we have become.

Technically, we all have access to the same amount of energy. But how our particular personalities are predisposed to use that energy differs. The challenge becomes how to best understand and manage our own energy dynamic. Luckily, we have help, another hallmark of humanity: shared information transcends time and place!

For example, Jung and his followers discerned and described that a human personality involves four main functions: how we receive vital information, how we process that information, how we evaluate that information, and how we use that information.

We receive information primarily either through our outer senses (seeing, hearing, touching, tasting, smelling) or through our inner senses (insight, ideas, intuition, inspiration, and integration). Ones preference is designated as either for Sensing or for Intuition.

I am an off the scale Intuitive. But this is measured on a line from one pole to its opposite because we are all capable of and require both for health and wholeness: it just depends on how much energy it takes to access and engage ones 'inferior' preference.

Where we process information determines whether we are Introverted or Extraverted. As a severe introvert, I go within to process. Extraverts prefer to bounce information off other people. But people drain rather than energize me, and while

I CAN preach, teach, parent, lead a parade, run a meeting or run a hospital ward, I then have to retreat inside. This is why I can get lost while driving home from church after attending to worship, coffee hour, and a couple of committee meetings.

Unfortunately, I was born into a culture that values extroversion and orients daily life around everyone always being on the go. I do much better in India and Latin America, where everyone takes an afternoon nap!

Do we evaluate the incoming information through head, or heart? For me, things have to make sense in order to feel right, which explains why I am attracted to Jung's ideas and have used them to understand both myself and others on personal and professional levels.

What we do with the received, processed, and evaluated information is reflected in how we organize our surrounding space and our place in it. I need to order the 'out there' because what's inside is often in a state of creative chaos. Thus I carefully plan and then carry out that plan rather than just go with the flow. Acting spontaneously puts me way outside my comfort zone, even though I dearly long to embrace that attribute.

None of these predispositions are either right or wrong, they just manifest as pronounced patterns in people, and give us clues as to who we truly are. I've found that being an introverted intuitive thinker who craves order is both a gift and a curse, and learning to work with both has co-created my authentic self. That Milt has the same preferences makes for an intriguing partnership: while we look at, interpret, and respond to the world so much the same that there's no need to explain or excuse ourselves or one another, our different genders and life experiences provide enough differences to keep us from 'co-signing our insanities.'

For it is what we do with our energy that defines us as unique personalities. Stars, planets, plants, and animals all radiate their particular identity. Each contains the 'self-organizing' dynamic, otherwise known as 'autopoiesis,' that points to the interior dimension of things, the inner capacity for self-actualization.

Now, of course, the cottonwood tree does this automatically, without self-awareness. Because as people we get to explore our own subjectivity through conscious participation, we can choose not to do so. It takes enormous courage to become your

Self, to acknowledge and accept your own uniqueness, your specific individuality, to stand up and say, "Here I am: someone who has never existed before and never will again!"

Individuation is the human process by which the universe continues to organize and differentiate itself in a time development sequence. When we fail to be faithful to what can only come into the universe through us, we thwart the whole evolutionary process.

Sin in this context is what you do to keep yourself from fulfilling your destiny; evil is when you prevent others from living out theirs, whether they be trees or creatures or other people.

Mercifully, we can mentally go back in time and reframe our past through the lens of current consciousness of strengths and weaknesses. For instance, while I cannot change the effect my shyness had on my band member classmate, I can reprogram my subsequent behavior.

Thus when I entered seminary half a lifetime after high school, I had the presence of mind to be clear with my classmates about my need for downtime, beg them not to take it personally when I didn't hang out with them in the kitchen between classes. In this way can I essentially redeem my future to be the best I can be with what I have to work with.

So why bother? I'll let Mumford have the final word here:

"It is these inner explorations, which date from man's first emergence from animalhood, that have made it possible to enlarge all the dimensions of being and crown mere existence with meaning. In this definite sense human history in its entirety, man's own voyage of self-discovery, is so far the climactic outcome of cosmic evolution."

Star Point Four

Spooking around in the dark so as not to disrupt Milt's sleep (the Einstein glow-in-the-dark face on his nightshirt finally faded), I reach for sweats and shoes and yoga mat. Two A.M. is way too early for anyone to be up and about, except if one wants to watch the Perseid Meteor Showers, which I do. For this is the best time of the year to catch sight of left-over debris still swirling around our planet from the explosion of our Mother Star. Positioning myself in the middle of the driveway so as to avoid the outside lights left on by neighbors for protection, I let my eyes adjust to the 'normal' Denver sky shine.

When I first see the streaks of lights shooting over my head through the heavens, I know in my bones these are grace-filled reminders of our cosmic beginnings, remnants of our origin story. And I am reminded of mathematical cosmologist Brian Swimme's observation:

> "To speak of the universe's origin is to bring to mind the great silent fire at the beginning of time. We can see the dawn of the universe because the light from its edges reaches us only now, after traveling twenty billion years to get here. Scientists have only just learned to see the fireball. The light has always been there, but the ability to respond to it required a tremendous development of the human senses. We can now see the beginnings of time. We are the first generation to live with an empirical view of the origin of the universe. We are the first humans to look into the night sky and see the birth of stars, the birth of galaxies, the birth of the cosmos as a whole. Our future as a species will be forged within this new story of the world."

But the familiar old story retains its power over me. "Let there be light" from studying Genesis back in my fundamentalist Christian Sunday School is hard to unlearn. More recently, it was the passage read by the first astronauts orbiting the moon and photographing the earth from space for the first time, even as they were living into a whole new story.

Since then quantum physicists have been expanding on Einstein:

"Let there be light! Is indeed a very profound statement. If you could ride a beam of light as an observer, all of space would shrink to a point, and all of time would collapse into an instant. In the reference frame of light, there is no space and time" (*Science and Spirit Magazine*).

What if, within the human brain, this single beam of light refracts, as through a prism, into a trinity of Eternal, Everyday, and Emergent Time?

Being fully present in the present moment, the now here, reveals the presence of the no where, the ineffable heaven that 'lies about us in our infancy' as Wordsworth poeticized.

In fact, the lost Gospel of Thomas claims Jesus taught that heaven is spread upon the earth but we do not see it.

Joseph Campbell notes that we indeed live in both the now here and the no where simultaneously: it's just a matter of where we place the 'w'. But actually moving into that no where realm is surprisingly complicated.

Being able to see the transcendent in the ordinary requires a shift in perspective that CAN happen in a once and for all conversion experience, but more often than not takes deliberate cultivation over time.

Often this requires smashing ones carefully constructed small self first, usually after midlife, so that the cosmic Self, ones own inner star, can connect with the transcendent light from the beginning of all time.

We know it is time to do this when the life we've constructed begins to fall apart, often through trauma or tragedy, and we find ourselves tottering on the edge of disaster, which, remember, means 'separated from the stars,'....and on the verge of a breakdown OR a breakthrough.

It is often only after giving up and giving in and giving over to something larger than the small ego self that we can come to know the Cosmic Self....the Divine Self...the Whole and Holy Self.

Mystics cultivate a path to this space through spiritual disciplines. And as Huston Smith points out, the mystics of all religious traditions report the very same experience of unity and inter-being.

It takes faith to risk being absorbed into the All. But it is there that we stand at the edge of Emergent Time, the opening of a creative new narrative for the whole world, that place where the future reaches into the present and shapes itself through the human imagination.

Yet the resistance is real! Who/where will I be if/when I become one with all that is here and now and ever was and ever will be?

For being in this new story means moving into a new state of being:

"In terms of quantum mechanics, you as an individual body are represented by a particular quantum state.

This includes all the elementary particles of your body. When you are intensely watching something, the photons of light reflected from that thing interact with your own elementary particles, and through this interaction, your quantum state is changed. Your particles are new in the sense that they have absorbed something from those other photons and entered a new state of being. You have become a brand new human being resonating with what you have absorbed" (Swimme).

This shift in perspective moves me beyond everything I've previously known about the natural world, especially trees. I knew that my life cycle depends upon the oxygen they 'breathe' out, and that their life cycle depends on the CO2 I breathe out. I also knew that we literally share the same molecular structure.

And trees have always been important to me figuratively as well.

For example, Joseph Campbell described the metaphoric Tree in the Garden of Eden as representing both the tree of the knowledge of good and evil of western traditions and the tree of eternal life honored by eastern traditions as both being limbs of the very same tree. And in my Transcendentalist tradition, Emerson claimed: "if you agree with me I may yet be wrong, but if the Elm tree says the same thing, I know I am right."

But knowing we are exchanging photons and being changed in the process changes everything. That cottonwood tree on the trail, gathering up the sun's light into its leaves that turn yellow in autumn and give it back as a glow, that tree shedding its twigs filled with stars, demands something of me. If I am radiating its essence, am I not also its mind, so to speak? What will that mean to how I live my ordinary human life?

"At a deep level, each living thing is implicated in every other. Ultimately, it is not individual species that evolve as much as all living systems connected interdependently within a coherent whole," according to Diarmund O'Murchu. And cell biologist Lynn Margulis has shown that 'natural selection' has more to do with cooperation than competition.

Consciously realizing that we live in an alive universe, and that life in our universe thrives not in isolationism but on the capacity to relate, also illustrates a Bantu proverb: "I am because we are."

Carl Sagan put it this way:

"We are the local embodiment of a Cosmos grown to self-awareness. We have begun to contemplate our origins: star stuff pondering the stars; organized assemblages of ten billion billion billion atoms considering the evolution of atoms; tracing the long journey by which, here at least, consciousness arose. Our loyalties are to the species—all the species—and the planet itself. WE speak for earth. Our obligation to survive (and thrive) is owed not just to ourselves but also to that Cosmos, ancient and vast, from which we have sprung."

As morning's light begins to brighten the sky, shutting off the sprinkling of stars and shutting out any hope of seeing more meteors showering our planet, I roll up my mat and move inside, assured that even while I can't see them, they're still there, with their message and mandate.

Star Point Five

Today the moon will eat the sun. Bit by bit, bite by bite, a chunk of earth will devour Earth's source of all light and life. Only an empty plate with a corona of luminous crumbs will be left.

Here in Denver, the moon won't get it all! A slice of sun will remain, pushed to the side of the plate like the rind of a melon.

Milt and I set up in his backyard, joining the coast-to-coast block party being broadcast by NASA, our publicly funded space agency.

When my brother's son and his son spent an overnight with us on their way to witness the totality up in Nebraska, they were joining the throng of humans clogging the interstate highway on a trek to have a first-hand experience of living in a cosmos!

We've opted to observe the cosmic do-si-do a few steps beyond our back door. From there we'll comfortably and consciously connect with the curiosity that took our species from caves to computers.

Of course, we first had to make a special trip to the science museum for our protective eclipse glasses, before they were gone from there too. But Milt has also constructed a box viewer and dug out an old colander for passive watching. A white poster board completes his preparation.

I drag out camp chairs and spread a blanket on the grass for reclining comfort: I fully intend to stay outside for the full three-hour duration of the eclipse. Water bottle, sunblock, hat, transition sunglasses, solar eclipse glasses, and umbrellas for some body shade should about do it.

I barely get settled and look up in time to glimpse the first nibble: a taste that's so tenuous we aren't sure it's truly happening.

But then a dark spot begins to grow, spreading from the top right edge of the sun down towards its center.

Soon I am alternating between staring through special lens glasses that block out everything but the fireball in the sky where the sun should be, and watching what Milt is up to with his ground experiments.

Surprises abound. Seeing the half-moon image of the sun reflected through the holes of the colander is as fun as expected. But the unexpected refraction of the sun's diminishing crescent through the leaves of a shade tree is a sheer delight.

Plus, Milt has duct-taped a pair of binoculars onto a tripod and pointed them at the propped up white poster board. What we see seems to be an extraterrestrial's head, its two bulging eyes changing with the sun. We nickname our space companion EC, for E-Clips.

As the sun slowly disappears, we're surprised when we realize that our transition prescription sunglasses are no longer darkening.

We are also surprised when it is not miserably hot so near noontime.

Milt runs inside to check the output of his solar panels as recorded on his computer: what should read 6.05 Kilowatts instead says .55 KW! I'll check mine back in Vegas later.

As the moon's shadow reaches its 93% fullness, we are both sitting still, simply staring. The sliver of light at the bottom of the sun hangs on by a fingernail in the dimming sky.

We hold our breaths, unable to imagine what's next.

The picture on Milt's new shirt from the science museum shows a wooly mammoth against the Denver skyline, a reverse half halo over its head. But how can that happen?

We keep staring while the rind of left-over light slides up the right side of the dish/disc of the sun until it looks just like the picture on his shirt.

While the sun begins reasserting its truth-claim on the sky, I dash inside to make sandwiches. Turning on the TV briefly, I catch a glimpse of the president squinting and pointing at the sun…without wearing his NASA issued glasses. I quickly switch channels, find an astronaut excitedly declaring that our country is on the verge of a new story.

Indeed, I believe we that are: one nation under the sun and the moon and the stars that have popped out in places experiencing totality.

After all, ours is the country that put the first person on the surface of the moon and began a new global narrative with one small step for man, one giant leap for mankind. Of course, everyone of my generation knows right where we were and what we were doing on July 20, 1969.

Milt was watching television at home, after helping to design the rocket that launched the historic mission.

I actually missed it. It wasn't until a day later that I got to listen, over transistor radio, to the otherworldly event.

As mission control monitored and calmly narrated the descent from the lunar module to the face of the moon, I was volunteering with a medical team in a rural Nicaragua. We'd just delivered a baby, placed it on its mother's belly, and were trying to shoo away the family rooster that had hopped up to have a look. Outside the nearly windowless house that lacked electricity and running water, my small daughter and her friends were drawing airplanes in the dirt, while the siblings of the newborn drew horses and chickens.

The disconnect was too great to be grasped. Even as I listened to the rebroadcasted event, I couldn't make sense of it. My mind was struggling between the Apollo mission underway above our heads and the human drama playing out around us in the earthly realm of his twin sister Artemis, the Greek goddess of childbirth.

Today, watching the moon slid off the disc of the sun, I can give thanks for the unexpected results of that narrative-changing mission.

The first photos of our planet from space changed everything! Our globe became one Earth, with no boundaries between countries: there were just continents and oceans and icecaps and rainforests and deserts and mountains: all interconnecting, interacting, interfacing.

In our country, inspired citizens organized the first Earth Day, lobbied the government to protect the air, water, and soil from further pollution, and began living more mindfully of the amazing planet we all call home.

I suddenly think to ask Milt, "why isn't the moon more like the earth if it's just a blasted off chunk of us?

He patiently explains how the moon's small size (even though it has just covered up the sun!) means that it has too little gravity to hold onto a protective atmosphere. And our moon does not have a protective magnetic field like the Earth. Without that protection from the sun's UV radiation and solar wind, life can't take hold.

OH! This is why ours is called the Goldilocks Planet! It is just right for what has come to pass here, and, as far as we know now, only here.

So why are we destroying it? Why are we humans messing with its atmosphere and thus changing its life-giving and life-sustaining climate?

As the blistering heat of full sun returns, I pull my chair into the shade. The cosmic event may be finished, but I'm not done yet.

Looking down at my arms, I return to the timeline that Milt and I constructed:

Since the Cambrian explosion of life at my elbow, there have been five mass extinctions. Four of these have been climate-change related.

The fifth extinction, caused by the asteroid that took out the dinosaurs, also became a climate event when the resulting debris altered the atmosphere and created a planet-wide winter.

A sixth mass extinction is already underway.

Staring down at my own hands, I see that within the wrinkle of the line at my wrist, modern humans walked out of Africa and landed on the moon.

In one hand is etched my own lifeline. The other holds the whole story of humanity, a species whose beliefs and behaviors are so radically changing the earth's atmosphere, biosphere, hydrosphere that we are bringing about the next mass extinction.

We are the first species to be able to look back over all the previous eons and reflect upon that awesome story. We are also the first species to be able to project into the future, see what we are doing, and figure out how to stop doing it before it's too late.

The choice is literally in our human hands.

Looking over the whole timeline, how am I to truly grasp that the emergence of each life form over those 3 billion years was a one-time only event, never to be replicated?

Death is the end of life; extinction is the end of birth.

Make no mistake: Life itself will regenerate into new species for niches that are unimaginable today. But the odds that there will be another permutation of the human are not promising.

Are WE to be a dead-end species?!

Are we humans so disconnected from the cosmic source of our being that we're endangering our becoming? How can it be that the human, the species through which the universe has become mindful of itself, is mindlessly creating the conditions of its own extinction?

While it appears that we are doomed to go the way of the dinosaurs, and the earth go on without us, as clergy I can't quite give up on the human species. Surely we haven't come this far to be just shucked off the planet by our folly!

Yet ours is the country creating most of the emissions that are warming the entire planet; ours is the only country that has dropped atom bombs and stockpiled enough weapons of mass destruction that if detonated would bring about a nuclear winter.

"Some say the world will end in fire, some say ice," wrote my hometown poet Robert Frost. Growing up, I assumed he was speaking metaphorically. Now, as an elder, I know how literally accurate this is.

Could today's shared solar eclipse be our much-needed national reset? Yet we the people are so disconnected from the universe that local parents complained to a local news station, demanding to know why the eclipse wasn't scheduled for when the kids were still home from school!

I actually get that! We humans have become so used to being in control of the natural world we forget that natural forces exist independent of our wishes. I found that out while taking flying lessons.

No amount of willpower would make up for miscalculating the headwind that would eat up more fuel than I had planned for. Having a fit in the cockpit at 5000 feet would not compensate for failing to climb up on a ladder with a ruler to physically measure the gas level in the wing tank before setting off on a solo cross-country trip. I learned not to fight natural forces, such as gravity, but to work with them.

In fact, it was these First Principles of physics that made human flight possible. Milt worked with these dynamics to send rockets to the moon and beyond. These immutable laws of the universe were laid down at the beginning of time and have been carried on through our Mother star and by our 'father' sun. We defy and deny their existence at our own peril.

Try to stretch a glide by pulling up on the nose without first adding power and I'll stall out the wings and make my Cessna drop like a rock. Dump too many greenhouse gasses into the atmosphere for too long and accept the consequences: severe hurricanes, acidic oceans, out of control wildfires, catastrophic floods, massive droughts, and species extinction.

The ultimate reality of nonrenewable natural resources and the growing evidence that human-caused heat-trapping emissions are irrevocably altering global temperatures demand a brand new frame for our beliefs and behaviors…IF our species is to continue to survive, let alone thrive.

"If you don't understand nature's laws now," warns Onondaga elder Orin Lyons, "You will soon."

Today's solar eclipse was a natural event so precisely predicted by science that it is being watched by millions of fellow citizens.

And, yes, the local schoolchildren were issued free solar eclipse glasses so they could go outdoors during class to be part of the national 'party.'

Meanwhile, on a remote hilltop close to the geographical center of the continent, my nephew and his son spread a blanket in a field of sunflowers to experience the cosmic event they have traveled so far to see, and will share with me a few days later:

"At 1% the world went grey. The temperature dropped and the wind picked up. Around us, the sunflowers started closing. At 100% we whipped off the glasses and saw perhaps the most surreal landscape I've ever seen. The sky was dark indigo and the sunset, rather than being in the west, we had a 360-degree sunset. A few stars had come out, in the middle a black hole poked through the sky surrounded by a ghostly corona. It was as if we were standing in the middle of a book cover from a science fiction novel.

I can see how ancient peoples thought the world was ending. Even being in the 'modern' world and knowing the science behind it, it was awesome to watch. Science only adds to the excitement and the mystery and the awe…rather than subtracts" (Dana Collins).

I am glad and grateful that the next generations get the connection between wonder and wondering. For awe IS the beginning of wisdom!

And science is simply the continuation of human consciousness as it grows in complexity!! I suddenly recall the words of an early writing mentor who was fond of quoting her Welch coal miner grandfather:

"You can have your airplanes, but you lose the wonder of the birds."

But I was beginning my flying lessons at that time, and countered with "that's not true for me…. rather, I am even more amazed that dinosaurs learned how to fly and that we humans have learned how to imitate them!"

Perhaps it is within this dance of wonder with wondering that the paradox of human potential moves us towards Joseph Campbell's vision:

"Each of us, whether we know it or not, is equally the Mind at Large, the laws of which are the law of not only of all minds but of all space as well. We are the children of this beautiful planet that we have seen photographed from the moon; we are its eyes and mind, it's seeing and thinking. And the earth, together with its sun, this light around which it flies like a moth, came forth, we are told, from a nebula, and that nebula, in turn from space. So we are the mind, ultimately, of space.

No wonder then that its laws and ours are the same, and that our depths are the depths of space.

Our [new story] therefore is to be of infinite space and its light, which is without as well as within. Like moths, we are caught in the spell of its allure, flying outward to the moon beyond, and also inward."

Reaching for a batch of cottonwood twigs retrieved when the backyard tree was taken down for safety this summer, I stare at their stars.

Have I been faithful to the message of their mandate?

While I can't know for certain, I do know that I shall keep on trying. And just declaring that intention can set vibrations of energy into motion and lead to a synchronicity on the quantum ('bundled' energy) level.

At least that was what a favorite seminary professor insisted would happen when you 'point your bright sword of intent'…i.e. focused your consciousness. And while her lectures about a butterfly flapping its wings on one side of the globe affecting the weather on the opposite side usually had me and a fellow classmate pounding the Berkeley pavement, trying to wrap our heads around quantum theory and non-localized energy, until we had walked the block so many times we were physically exhausted as well as mentally frustrated, I know she was right about focused intention.

So I'll take to taping Milt's tiny slices of cottonwood twigs into birthday wishes and Christmas greetings.

And just as others carry around rosary and prayer beads, I'll make it a point to carry a fistful of stars in my pocket or purse, so that I may step out of my shyness and stop passersby with:

"May I show you something?"

Addendum

For Reader Participation

Star Point One

While I can't personally hand you a twig of cottonwood with its perfect little star, I can invite you too to commune with the cosmos through the spiritual discipline of writing, which is simply 'praying with a pen,' as Huston Smith put it.

First comes centering yourself in the great mystery of the cosmos, becoming grounded in the graced amazement that everything contains star stuff, not just the cottonwood tree.

So select a subject from the natural world that 'speaks' to you: a feather, a leaf, a seashell. What wants to become your organizing metaphor for this cosmic-self journey? You may decide to arrange your treasure/s around a candle that symbolizes the light from the beginning of Time that still abounds.

A lit candle recreates the warmth and light that began with the Big Bang and continued on through our local star and called forth all life on our planet.

Without our sun's 4.5 billion years of faithful, enduring, and persistent heat and light, we would not exist.

A single seashell can signify water, the primary material from which all life flows, as 'in the beginning,' i.e. the 'deep' of many creation myths. Plus personally; each of us gushed forth from the watery womb that cradled our beginning. And the wonder of this truth connects us with all people, past and present and future, from the ancient Hindus who spoke of the World Egg floating upon the primal ocean, from which the entire cosmos was born. The water that washed Jesus' feet was still around to baptize your great grandmother, and will still be here to dedicate your great grandchildren.

Can you feel that ancient universal element making up 90% of your body, flowing through you, transporting energy throughout your system, and all life systems? For what are we humans, Loren Eiseley asks, but a clever way that water has found for moving beyond the reach of ponds?

A feather can stand for the air and make you aware that every molecule of O_2 you inhale has been circumnavigating the planet for billions of years, and been shared by the famous and infamous. When you breathe in are you aware of the oxygen upon which your cells depend for the continuance of your very life and that the carbon dioxide you breathe out is what plants need for the continuance of their life cycle?

A single leaf can symbolize the plants that concentrated the Sun's energy into a form needed to drive your high metabolic rate. Or as Eiseley put it: "The human brain, so frail, so perishable, so full of inexhaustible dreams and hungers, burns by power of the leaf."

Then pick up your pen:
we are all creating the new narrative.

At the end of each section you will find writing prompts to reflect upon as you compose your piece of the unfolding story.

Begin with writing about which of these images (leaf, shell, feather, etc.) speaks most strongly to and through you. For your writing hand is but a sequence of earth's evolutionary journey, our mother star's memory poured into your flesh, fiber, and bone as your own DNA moved from fin to flipper to the finger steadying your pen.

Thus what wants to come through you really comes from the Cosmos itself as star stuff reflecting upon star stuff.

Star Point Two

While you were growing up in your 'second womb,' your family of origin, did you fit in like a fish in water, just going with the flow of things? Or were you more like a budding amphibian, drawn towards something more? The toddler saying NO to everything becomes the teen (just a tall toddler?) pushing against the norms of the family as its 'womb' becomes too small, but then facing the angst of alienation. Questioning things as they are sets up a sense of not fitting in that CAN lead to self-destructive behaviors and choices.

BUT that stage of feeling like a fish out of water can be the liminal space in which you begin to differentiate who you are from everyone and everything else. For to Be is to embrace and be embraced by the cosmic dynamic that creates the diversity, complexity, variation, disparity, and multiplicity that keeps the process of evolution going.

A youngster once asked her teacher: "what is it like to be a grown-up? Do you feel tall, fat, and all finished?"

The bad news is this: you are not all finished,

The good news is this: you are not all finished.

Because our human 'programming' is so open-ended, we get to remake ourselves at every stage of our lives. What might you finally need to say a great NO to before you can move forward into what is emerging through you as our species continues to 'progress?'

Star Point Three

Often we don't know who we really are until and unless we rub someone the wrong way. While at first this can feel like a curse, it is really a gift. Of course, criticism says as much (if not more) about that other person as it does about you, for when you point your finger at someone, three more are pointing right back at you.

But this also helps us identify in them what we can't yet see or accept in ourselves. In fact, how we affect others makes us conscious of what is usually unconscious in ourselves. Why have we repressed this piece of ourselves, kept it hidden in the shadows? What's down there in our Shadow can break through at unexpected times and in inappropriate behavior, and sometimes presents itself in dreams and nightmares.

Yet your Shadow may contain a true treasure just waiting to be retrieved and polished up and presented to the world. But this means acknowledging and accepting that part of yourself you don't want to own.

Begin this process by admitting to one thing you have been criticized for and/or you feel guilty about. Can you stop 'beating yourself up' over it and begin to accept it as a gift instead of a curse?

If so, then it is time to say the great YES to who you really are, your own cosmic dynamic known as subjectivity, the self organizing principle that points to the interior dimension of things, the inner capacity for self-manifestation, self-organization, identity, interiority, and unseen shaping.

For then you will become YOU, and not a second hand version of some revered role model, such as Moses, Jesus, Mohammed, Buddha, or Krishna.

They have already manifested their part of the great arc of the story.

Who/where are YOU?!

Star Point Four

How amazing is this, that during the annual season of darkness, humans celebrate light? As the sun's light diminishes with the winter solstice, festivals of light abound: Christmas, Hanukkah, Diwali, Kwanzaa all build upon the rites and rituals first laid down by ancient humans in an effort to entice the Sun to come back with its life-giving light.

We humans seek enlightenment, both personally and collectively, as in The Enlightenment, that Age of Reason that ushered in scientific methodology along with the ideals of liberty, progress, tolerance, and constitutional government. Celebrated as 'the Century of Lights,' this era continues to work into and through current human awareness.

Many of us grew up singing 'this little light of mine, I'm gonna let it shine' in Sunday school or scouts or youth groups. But what if this little light of mine isn't really mine? What if the light that became my life and provides enough energy within my body to power a 100-watt light bulb belongs to the great trajectory of Light from the beginning of Time?

Let me invite you to ponder a question posed by Parker Palmer:

"Is the life I am living the same as the life that wants to live in me?"

Star Point Five

Ah, the stars: we wonder what they are, we wish upon them, we follow them, and we thank our lucky ones…all as if they are somewhere 'out there,' beyond our grasp, rather than within and among us here and now.

In fact, we 'commune' with them all the time.

But we usually don't become conscious of this until that reality commands our attention. For instance, we 'forget' that our blood circulates the iron forged in the stars until we become anemic and need to replenish our personal supply. We ignore that our bones are made up of star-composed calcium until osteoporosis makes them brittle and breakable and we take action to rebuild them. Of course we do load up on carbs for our carbon-rich muscles before running a marathon. And we know we must drink lots of fluids to fend off dehydration in the heat.

Perhaps we could be more conscious of these freely bestowed gifts as we partake of the 'body' and 'blood' of the cosmos. And say thank you.

For grace-gracias-gratitude is a perfect lens through which to perceive, and receive, the star stuff you come from…..along with your place in the cosmos. For "you are a child of the universe, no less than the trees and the stars," as Max Ehrmann wrote in his Desiderata.

Gratitude can help you reclaim the past, reprogram the present, and redeem the future. Looking over the body of work you have created while responding to these writing prompts, you can review your past for what went right rather than what went wrong. For often your greatest failure is the opening through which your authentic self comes forth.

You can be fully present in the present moment and appreciate even the most 'mundane' cosmic gifts, such as feeling the sun's warmth on your skin, sipping clean water, breathing fresh air, consuming good food. And if the water is contaminated, the air polluted, the foodstuff poisoned, you can wonder why and decide what you will do about it.

You can approach the future by changing old habits out for new ones, thus becoming proactive rather than reactive...as an adult of the universe.

With your 'right' to be here comes response/ability.

Within the context of the cosmos, what is the life that wants to live in and through you? What is the meaning of your time in-between being 'from ashes to ashes, dust to dust, star stuff to star stuff?'

Your true mission can be found at the place where your individual gifts, interests, and affinities intersect with what needs doing in the world. It is in that space that your life-star shines through, proclaiming to all others:

"May I show you something?"

Some of Our Cherished Sources of Illumination

Berry, Thomas. *The Dream of the Earth.* San Francisco: Sierra Club Books, 1988.

Berry, Thomas. *The Great Work.* New York: Bell Tower, 1999.

Campbell, Joseph. *Myths to Live By.* New York: Viking Press, 1972.

Dowd, Michael. *Thank God for Evolution.* San Francisco: Council Oak Books, 2007.

Eiseley, Loren. *The Immense Universe.* New York: Vantage Books, 1946.

Goodenough, Ursula. *The Sacred Depths of Nature.* NY: Oxford University Press, 1998.

Grinspoon, David. *Earth In Human Hands.* NY: Grand Central Publishing, 2016.

Keirsey, David and Marilyn Bates. *Please Understand Me.* Del Mar, CA: Prometheus Nemesis Books, 1978.

Mumford, Lewis. *The Myth of the Machine.* NY: Harcourt, Brace, & Jovanovich, 1970.

O'Murchu, Diarmuid. *Evolutionary Faith.* New York: Orbis Books, 2003.

O'Murchu, Diarmuid. *Quantum Theology.* New York: Crossroad Publishing, 2000.

Rue, Loyal. *Everyone's Story.* Albany, NY: SUNY Press, 1999.

Sagan, Carl. *Cosmos.* New York: Random House, 1980.

Shubin, Neil. *Your Inner Fish,* New York: Pantheon Books, 2008

Swimme, Brian. *The Hidden Heart of the Cosmos.* NY: Orbis Books, 1996.

Swimme, Brian and Thomas Berry. *The Universe Story.* San Francisco: Harper, 1992.

About the Authors

The star stuff that makes up the human GAIL COLLINS-RANADIVE has taken on a series of manifestations that include daughter, big sister, student, friend, reader, wife, mother, psychiatric nurse, military spouse, children's book author, private pilot, feminist, peace activist, poet, environmentalist, parish minister, grandmother, life partner, published author of 8 non-fiction books, and is currently engaged in trying to prevent catastrophic climate change and fend off the next mass extinction, at human hands.

The star stuff comprising her partner, the human MILT HETRICK, has manifested as in the form of son, big brother, farm boy, student, college graduate, licensed professional engineer/physicist, husband, father, stepfather, life partner, pragmatic iconoclast, non-violent conflict management advocate, environmental advocate, sustainable living advocate, practicing geek and web site developer, and amateur writer.

LITTLE
BOUND BOOKS
SMALL BOOKS, BIG IMPACT

The Little Bound Books Essay Series
Personal. Poignant. Powerful.

WWW.HOMEBOUNDPUBLICATIONS.COM

HOMEBOUND PUBLICATIONS

Ensuring that the mainstream isn't the only stream.

At Homebound Publications, we publish books written by independent voices for independent minds. Our books focus on a return to simplicity and balance, connection to the earth and each other, and the search for meaning and authenticity. Founded in 2011, Homebound Publications is one of the rising independent publishers in the country. Collectively through our imprints, we publish between fifteen to twenty offerings each year. Our authors have received dozens of awards, including: *Foreword Reviews'* Book of the Year, Nautilus Book Award, Benjamin Franklin Book Awards, and Saltire Literary Awards. Highly-respected among bookstores, readers and authors alike, Homebound Publications has a proven devotion to quality, originality and integrity.

We are a small press with big ideas. As an independent publisher we strive to ensure that the mainstream is not the only stream. It is our intention at Homebound Publications to preserve contemplative storytelling. We publish full-length introspective works of creative non-fiction as well as essay collections, travel writing, poetry, and novels. In all our titles, our intention is to introduce new perspectives that will directly aid humankind in the trials we face at present as a global village.

WWW.HOMEBOUNDPUBLICATIONS.COM

OCT - 2018

CPSIA information can be obtained
at www.ICGtesting.com
Printed in the USA
LVHW03s0217170618
580549LV00002B/2/P